SPARKS OF LIFE

Chemical Elements that Make Life Possible

SODIUM

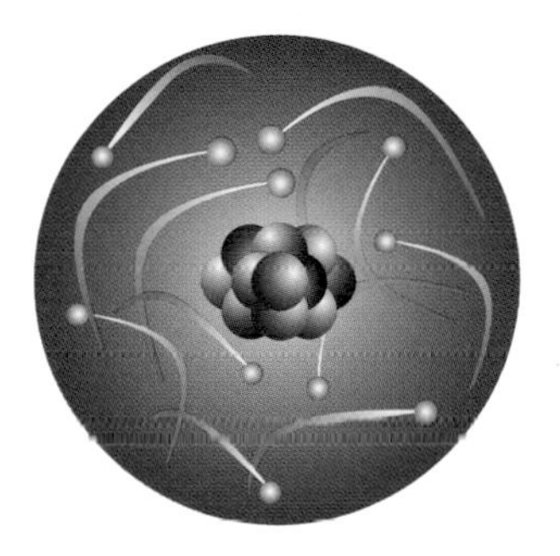

by

Jean F. Blashfield

Austin, Texas

Special thanks to our technical consultant,
Jeanne Hamers, Ph.D.,
formerly with the Institute of Chemical Education,
Madison, Wisconsin

Development: Books Two, Delavan, Wisconsin
Graphics: Krueger Graphics, Janesville, Wisconsin
Interior Design: Peg Esposito
Photo Research and Indexing: Margie Benson

Raintree Steck-Vaughn Publisher's Staff:
Publishing Director: Walter Kossmann
Design Manager: Joyce Spicer
Project Editor: Frank Tarsitano
Electronic Production: Scott Melcer

Library of Congress Cataloging-in-Publication Data:
Blashfield, Jean F.
Sodium / by Jean F. Blashfield.
p. cm. — (Sparks of life)
Includes bibliographical references (p. -) and index.
Summary: Discusses the origin, discovery, special characteristics, and uses of the sixth most abundant element in the Earth's crust.
ISBN 0-8172-5042-5
1. Sodium — Juvenile literature. [1. Sodium.] I. Title. II. Series: Blashfield, Jean F. Sparks of life.
QD181.N2B57 1999 98-15236
546' .382 — dc21 CIP
AC

Printed and bound in the United States.
1 2 3 4 5 6 7 8 9 LB 03 02 01 00 99 98

PHOTO CREDITS: ©Walt Anderson 23; Ball-Foster 45; The Bettmann Archive 36; P.K.-Chen and Yerkes Observatory cover, 54; Courtesy of Church & Dwight Co., Inc. 47; Culligan International Company 42; James Holmes/Reed Nurse/Science Photo Library 34; ©Bruce Iverson 25; JLM Visuals cover, 16, 17, 19, 27; Dr. P. Marazzi/Science Photo Library 32; NASA 21; ©John Hyde/Pacific Stock cover; ©Bruce Roberts/Photo Researchers 43; Science Photo Library 10; ©Michael Smutko 1998 52; Photo courtesy of U.S. Borax, Inc. 51; ©Charles D. Winters/Photo Researchers 13, 28; Wisconsin Paper Council 49.

CONTENTS

Periodic Table of the Elements

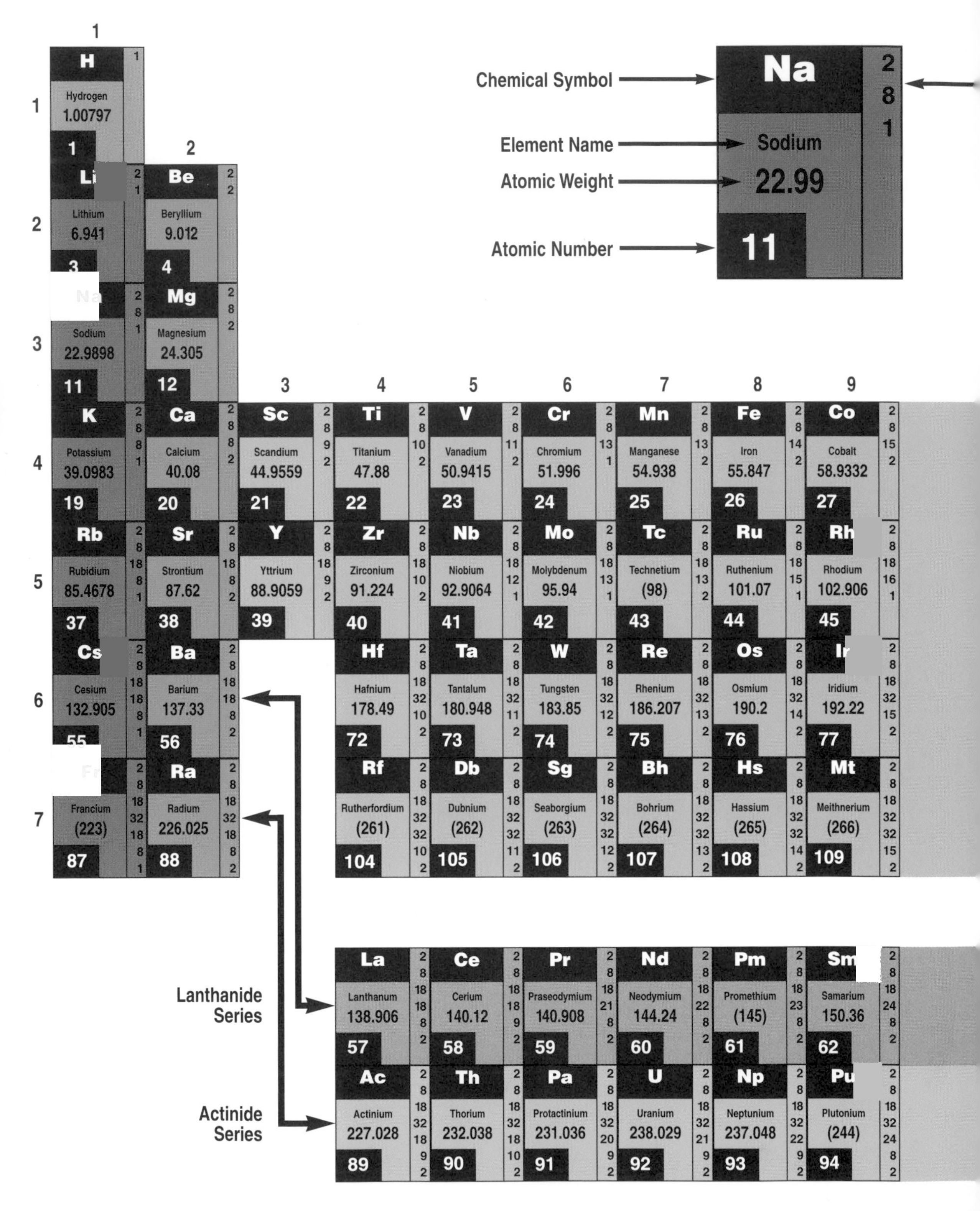

Number of electrons in each shell,
beginning with the K shell, top.

See next page for explanations.

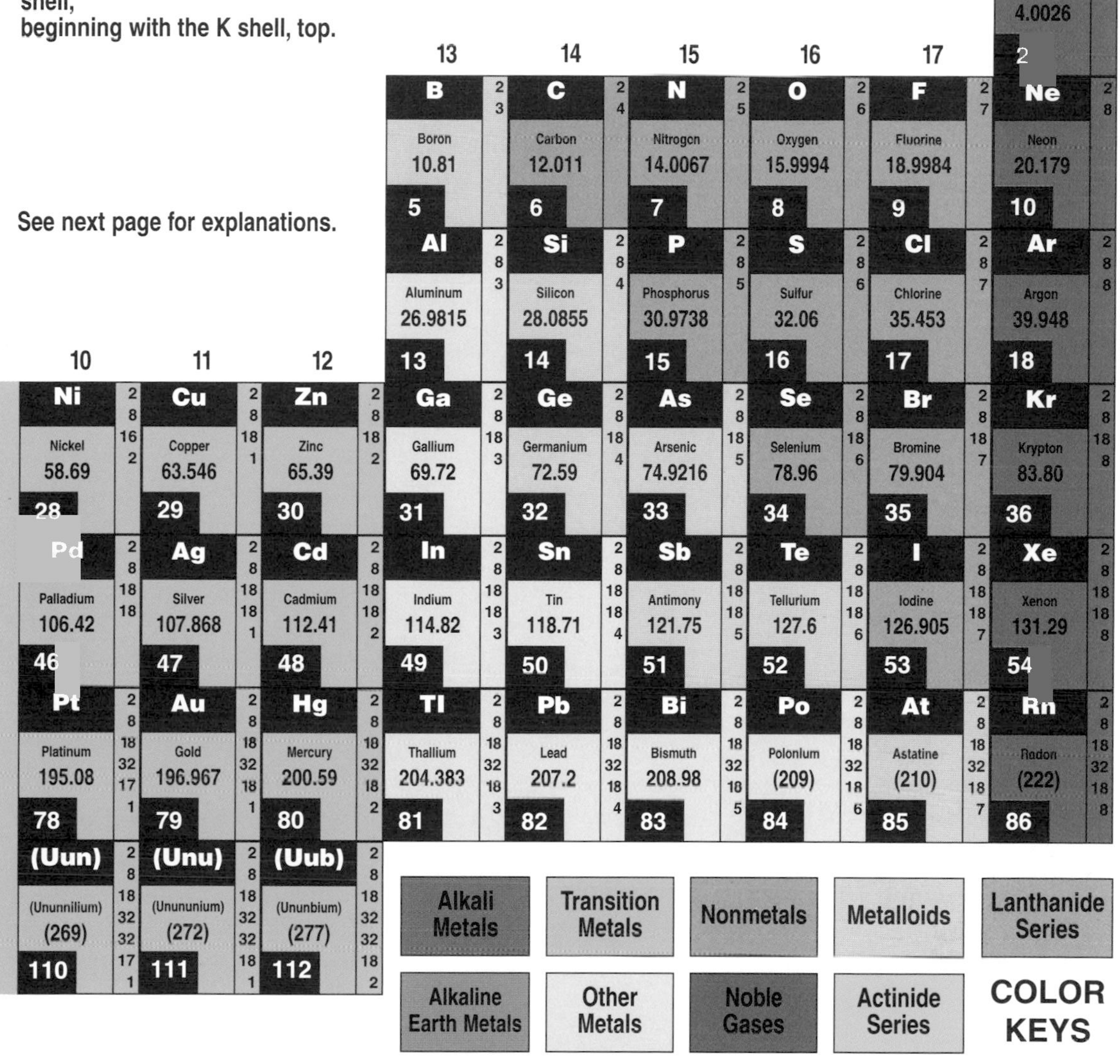

Symbol	Name	Atomic weight	Atomic number	Electrons per shell
Eu	Europium	151.96	63	2 8 18 25 8 2
Gd	Gadolinium	157.25	64	2 8 18 25 9 2
Tb	Terbium	158.925	65	2 8 18 27 8 2
Dy	Dysprosium	162.50	66	2 8 18 28 8 2
Ho	Holmium	164.93	67	2 8 18 29 8 2
Er	Erbium	167.26	68	2 8 18 30 8 2
Tm	Thulium	168.934	69	2 8 18 31 8 2
Yb	Ytterbium	173.04	70	2 8 18 32 8 2
Lu	Lutetium	174.967	71	2 8 18 32 9 2
Am	Americium	(243)	95	2 8 18 32 25 8 2
Cm	Curium	(247)	96	2 8 18 32 25 9 2
Bk	Berkelium	(247)	97	2 8 18 32 26 9 2
Cf	Californium	(251)	98	2 8 18 32 28 8 2
Es	Einsteinium	(254)	99	2 8 18 32 29 8 2
Fm	Fermium	(257)	100	2 8 18 32 30 8 2
Md	Mendelevium	(258)	101	2 8 18 32 31 8 2
No	Nobelium	(259)	102	2 8 18 32 32 8 2
Lr	Lawrencium	(260)	103	2 8 18 32 32 9 2

A Guide to the Periodic Table

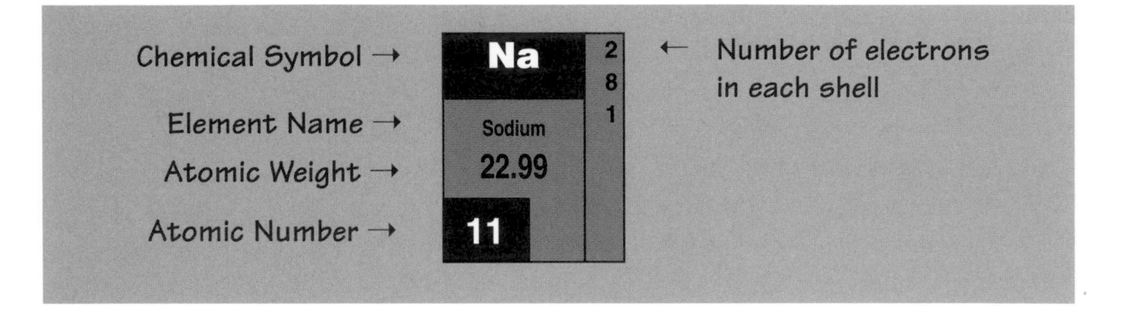

Symbol = an abbreviation of an element name, agreed on by members of the International Union of Pure and Applied Chemistry. The idea to use symbols was started by a Swedish chemist, Jöns Jakob Berzelius, about 1814. Note that the elements with numbers 110, 111, and 112, which were "discovered" in 1996, have not yet been given official names.

Atomic number = the number of protons (particles with a positive electrical charge) in the nucleus of an atom of an element; also equal to the number of electrons (particles with a negative electrical charge) found in the shells, or rings, of an atom that does not have an electrical charge.

Atomic weight = the weight of an element compared to a standard element, carbon. When the Periodic Table was first developed, hydrogen was used as the standard. It was given an atomic weight of 1, but that created some difficulties, and in 1962, the standard was changed to carbon-12, which is the most common form of the element carbon, with an atomic weight of 12.

The Periodic Table on pages 4 and 5 shows the atomic weight of carbon as 12.011 because an atomic weight is an average of the weights, or masses, of all the different naturally occurring forms of an atom. Each form, called an isotope, has a different number of neutrons (uncharged particles) in the nucleus. Most elements have several isotopes, but chemists assume that any two samples of an element are made up of the same mixture of isotopes and thus have the same mass, or weight.

Electron shells = regions surrounding the nucleus of an atom in which the electrons move. Historically, electron shells have been described as orbits similar to a planet's orbit. But actually they are whole areas with a specific range of energy levels, in which certain electrons vibrate and move around. The shell closest to the nucleus, the K shell, can contain only 2 electrons. The K shell has the lowest energy level, and it is very hard to break its electrons away. The second shell, L, can contain only 8 electrons. Others can contain up to 32 electrons. The outer shell, in which chemical reactions occur, is called the valence shell.

Periods = horizontal rows of elements in the Periodic Table. A period contains all the elements with the same number of orbital shells of electrons. Note that the actinide and lanthanide (or rare earth) elements shown in rows below the main table really belong within the table, but it is not regarded as practical to print such a wide table as would be required.

Groups = vertical columns of elements in the periodic table; also called families. A group contains all elements that naturally have the same number of electrons in the outermost shell or orbital of the atom. Elements in a group tend to behave in similar ways.

Group 1 = alkali metals: very reactive and never found in nature in their pure forms. Bright, soft metals, they have one valence electron and, like other metals, conduct both electricity and heat.

Group 2 = alkaline earth metals: also very reactive and thus do not occur in pure form in nature. Harder and denser than alkali metals, they have two valence electrons that easily combine with other chemicals.

Groups 3–12 = transition metals: the great mass of metals, with a variable number of electrons; can exist in pure form.

Groups 13–17 = metals, metalloids, and nonmetals. Metalloids possess some characteristics of metals and some of nonmetals. Unlike metals and metalloids, nonmetals do not conduct electricity

Group 18 = noble, or rare, gases: in general, these nonmetallic gaseous elements do not react with other elements because their valence shells are full.

THE ELUSIVE METAL

Common table salt was one of the first chemicals used by human beings. Our ancestors learned early that their bodies needed it to survive. They also found that this natural white crystal, which could be found in deposits all over the world, could preserve food and make it taste better. Long before we had refrigeration, we knew that food soaked in salt would last a long time.

The white crystals that we call table salt are only one of the important substances that include the chemical element known as sodium. The chemical symbol for sodium is Na, and it is atomic number 11 on the Periodic Table of the Elements.

Sodium is the sixth most abundant element in the Earth's crust. About

2.6 percent of the Earth's crust is sodium. The most common sodium mineral is rock salt, which is sodium chloride, NaCl. (Chlorine is Cl, element #17.) When the waters of ancient oceans evaporated, the sodium chloride already dissolved in the water crystallized to form huge deposits of rock salt, also called halite. Sodium also occurs in many other minerals.

Sodium compounds found naturally in the Earth's crust came originally from molten volcanic rock. Such rocks are called igneous rocks. When igneous rocks are exposed to the weather, they begin to wear away. The sodium from these rocks dissolves and enters rivers and lakes, eventually ending up in the oceans.

The Hunt for Understanding

The isolation of sodium as an element is closely tied to the isolation of the element potassium (K, element #19). The same man, English chemist and popular science lecturer Humphry Davy, was responsible for the isolation of both elements, using the same process, and within a few days of each other in 1808.

Scientists had been trying for a long time to analyze the substance they called potash. Though the name is now used for almost any potassium-containing mineral, they were referring to the white substance that was left when wood ashes were boiled in water. The experimenters also called potash vegetable alkali because it behaved like a mineral that they called both alkali and soda ash.

Until the eighteenth century, chemists made no distinction between potash and soda ash. Both of them were caustic, which means that they ate away, or corroded, other substances. Also, both potash and soda ash were useful in making glass. The chemists knew there must be some difference between the two substances, but they could not figure out what it was.

In 1794, Alessandro Volta, an inquisitive Italian count, succeeded in making the world's first battery. The brilliant Humphry

Davy became fascinated by Volta's electric battery, especially after he learned that electricity from a battery had been used to break water down into its basic elements of oxygen (O, element #8) and hydrogen (H, #1). Davy wondered whether the same process, called electrolysis, could be used to separate potash into its component elements.

Davy's Electrical Element-Isolating Device

Eager to test his idea, Humphry Davy made his own electrolysis apparatus. Electrolysis equipment consists of bars of two different metals attached to a battery. The bars, called electrodes, are submerged in a solution or fluid form of the substance being studied. When the battery is connected to the bars, electricity flows through the fluid.

The fluid being electrolyzed (called the electrolyte) breaks up into separate ions, which are atoms or groups of atoms with an electrical charge. Some of the ions have extra electrons, giving them a negative charge. Negative ions are drawn to the positive electrode, or anode, where they lose the extra electrons, becoming uncharged atoms of an element. Other ions have a positive electrical charge and are drawn to the negative electrode, or cathode. At the cathode, they take on electrons, also losing their electrical charge.

Humphry Davy

The first time Davy tried his experiment, he used potash dissolved in water, but he succeeded only in separating the oxygen and hydrogen ions that make up water. The potash was left untouched.

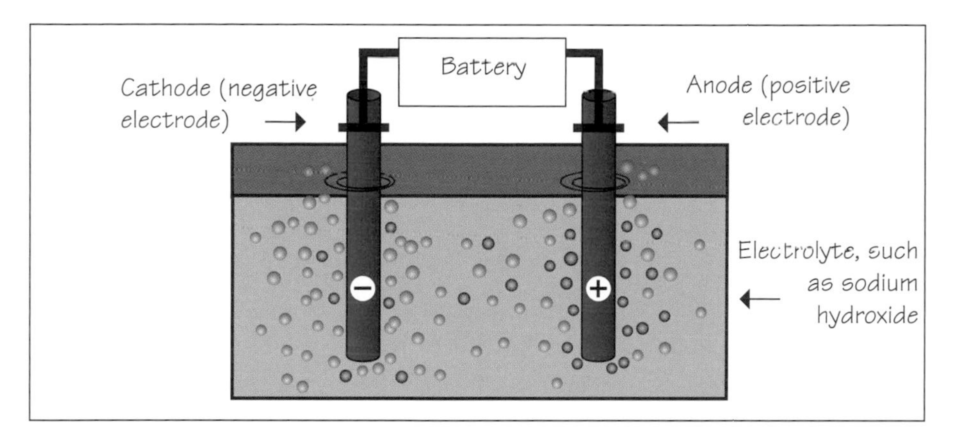

The simple electrolysis equipment used by Humphry Davy to separate sodium from other elements in such compounds as sodium hydroxide

Then, in October 1807, the young scientist tried his experiment once again. This time using molten, or liquid, potash, which he had made by heating the potash until it became a thick fluid. He was delighted when silvery metallic droplets formed on the cathode. He called the metal potassium, from the word "potash."

Davy then tried to isolate sodium in an electrolysis experiment using sodium hydroxide, NaOH. The experiment didn't work until he made the most powerful battery produced up to that time. With enough electricity, Davy succeeded in isolating a new metal, which he named sodium. Sodium's chemical symbol, Na (for *natrium*), is derived from the word *natron,* which was the name given to the salt deposits in ancient Egypt.

The sodium produced by Davy was not chemically pure. Obtaining pure sodium was one of the achievements of Robert Wilhelm Bunsen, the inventor of the Bunsen burner. About 1860, this German chemist was the first to obtain pure samples of several elements including sodium, lithium (Li, element #3), calcium (Ca, #20), and barium (Ba, #56).

Hiss and Fizz

Sodium and the other alkali metals (elements in Group 1—sometimes called IA—of the Periodic Table) are soft. Pure metallic sodium can be cut with a table knife. When cut, it looks white and silvery for a brief moment, but then it reacts with oxygen in air and loses its shine.

Alkali metals also float in water. However, if you try to float a chunk of sodium in water, it fizzes and hisses and gives off flames. Water molecules are made up of oxygen and hydrogen atoms. The sodium is reacting with the oxygen in water, causing it to release pure hydrogen gas.

Pure sodium has virtually no uses unless it is kept in an environment without air or water. If air or water is present, the metal reacts instantaneously by oxidizing—giving up the single electron in its outer, or valence, electron shell. Pure sodium is normally kept under a nonreacting blanket of nitrogen (N, element #7) gas or under kerosene in an airtight bottle.

In the presence of just a little oxygen, pure sodium reacts—violently—by forming sodium oxide, Na_2O. With just a little additional air, sodium oxide takes on another oxygen atom and turns into sodium peroxide, Na_2O_2. Pure sodium exposed to an abundance of air, however, turns directly to sodium peroxide. Sodium peroxide is an efficient bleach because it readily oxidizes chemicals that give color to many substances.

The Important Electron Shell

The atomic number 11 given to sodium means that it has eleven protons (positively charged particles) in its center, or nucleus. They are balanced by eleven electrons (negatively charged particles) that move in regions surrounding the nucleus

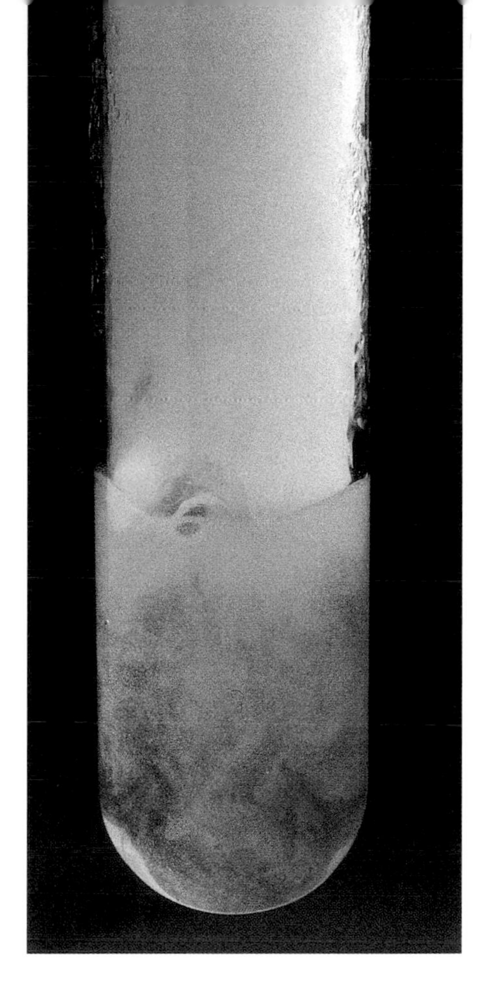

Bits of pure sodium react instantly to the presence of oxygen in air and water by bursting into flame.

called shells. The additional eleven particles in the nucleus are called neutrons and have no electric charge.

All elements except hydrogen have two electrons in the first, or inner, shell. Hydrogen has only one. Elements 3 through 10 (lithium through neon) have a progression of one through a maximum of eight electrons in the second shell, thus giving these a total electron count—and an atomic number—of 3 through 10. Sodium has an eleventh electron, which begins its third shell. Like the second shell, the third shell can contain a maximum of eight electrons.

Sodium's eleventh electron may reside in its third shell, but it doesn't stay there long. Having only one electron in its outer shell makes sodium a very reactive element.

Other elements in Group 1 of the Periodic Table are also highly reactive. They include lithium, potassium, and rubidium (Rb, element #37), cesium (Cs, #55), and francium (Fr, #87). Hydrogen is sometimes put at the head of the Group 1 elements because it too has only one electron in its single electron shell. However, it is not an alkali metal and does not react chemically like a Group 1 metal.

The Ion

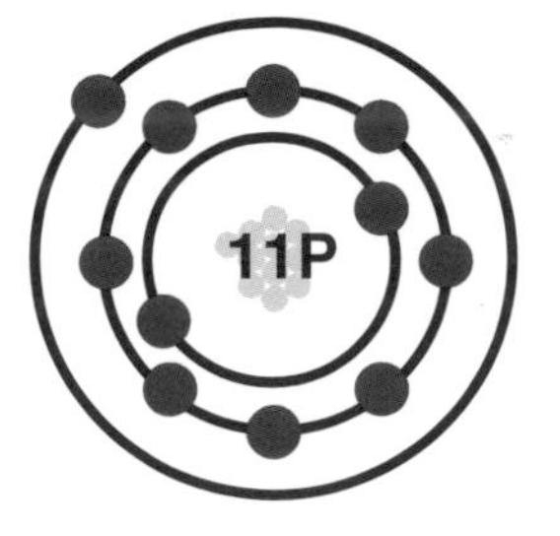

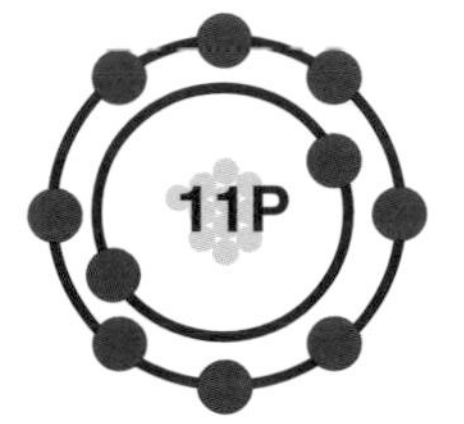

As soon as a sodium atom lets go of the single electron in its outer shell, the sodium atom turns into a positive ion, which is written Na+. It has a positive electrical charge because there is no longer a negative electron to balance the eleventh positive proton in the nucleus.

Because sodium reacts readily with oxygen to form sodium oxide, Na_2O, that compound is a common form of sodium in Earth's crust. However, flowing water from rain or rivers breaks the sodium oxide into ions. The oxygen ions go into the air, and the sodium ions remain in the water.

Seawater contains 10,500 parts of sodium ions per 1 million parts of water. The sodium ions (Na^+) are the second most abundant kind of ion in the sea, after negatively charged chloride ions (Cl^-). Like sodium ions, chloride ions also come from rocks, though they are rocks that contain chlorine. These positive and negative ions join to form electrically neutral salt, or sodium chloride, NaCl.

Salt is very important to human nutrition. Since prehistoric times, people have recognized that some salt was important to health. In fact, it was in trading supplies of salt for other products that much of the early world was explored and settled. Our word "salary" came from the ancient Roman word for the pay that soldiers received in order to buy their salt.

Animal life probably developed in the sea, and the sodium ions common to saltwater are an important part of the blood of all animals. Sodium is not essential to plant life. Except for seaweeds and some trees, such as mangroves, most plants are harmed by salt. Animal life needs sodium ions, but plant life can be damaged by sodium.

The confusion between potassium and sodium didn't stop causing problems once the elements were isolated as elements. In chemical laboratories, it can sometimes be difficult to distinguish between the two.

Ions of sodium add a bright yellow color to a flame. That is one way to test for the presence of sodium in a substance. Potassium burns with a pale purple color. Sometimes these elements are found together. When that happens, the yellow of sodium flame hides the pale purple flame of the potassium. However, if the flame is viewed through a blue glass filter, the blue blocks out the yellow, revealing the pale purple.

The Isotopes

Sodium is present in the sun and stars. It's even found in the atmosphere surrounding Earth. Sodium atoms exist in a thin layer of our planet's atmosphere starting about 70 kilometers (45 miles) up. They probably help cause the faint luminescence, or light, that exists in the night sky. Sodium has also been identified as existing in the core of comets.

Many elements exist in several forms, called isotopes, based on the number of neutrons, or neutral particles, in the nucleus. The word isotope means "same place." The different kinds of atoms of an element with varying numbers of neutrons all occupy the "same place" in the Periodic Table. The only stable isotope of sodium is common sodium-23, which has twelve neutrons and eleven protons in its nucleus.

Sodium-23 is stable, meaning its nucleus does not readily change. The other sodium isotopes are radioactive. Their nuclei give off, or emit, various particles until—over a long period of time—the atoms change into a more stable form.

Sodium Minerals and Rocks

Feldspars are igneous rocks that formed as molten material inside the earth and then cooled when they reached the surface. Feldspars consist of a framework of silicon (Si, element #14), oxygen, and aluminum (Al, #13) atoms, but they also contain other elements, such as sodium, potassium, or calcium.

When feldspars reach the surface of the earth, they immediately begin to weather, or wear away from rain and wind and temperature changes. This process puts sodium ions from the sodium-rich feldspars in the rainwater, which flows to the rivers and eventually into the ocean.

Rocks that are rich in sodium and potassium are called alkaline rocks because the rocks contain an abundance of minerals made from alkali metals. One of the most common sodium-bearing minerals is borax, or sodium tetraborate ($Na_2B_4O_7 \cdot 10H_2O$). Perhaps the most beautiful is jadeite ($NaAlSi_2O_6$). Among other sodium compounds found in deposits are sodium carbonate (Na_2CO_3) and sodium nitrate ($NaNO_3$), which is also known by the name saltpeter.

A sample of feldspar rock. Feldspars are among the sources of the sodium that reaches the oceans.

OF SALTS AND SEAS

Probably the first chemical name we all learned is sodium chloride, which is the scientific name for table salt. If you place a chunk of sodium metal in a container of chlorine gas, the metal instantly flares up as if it will burn away. But when the flames disappear, what is left behind in the container? White crystals that look like salt, taste like salt, and are salt!

The flame is the outward sign that sodium atoms are losing their valence electrons to chlorine atoms. That creates positive sodium ions and negative chloride ions. Ions with positive charges (cations) combine with ions with negative charges (anions) in what is called an ionic bond. Because of the way the sodium and chloride ions unite with each

Rock salt breaks along flat planes because it has a cubic crystalline structure.

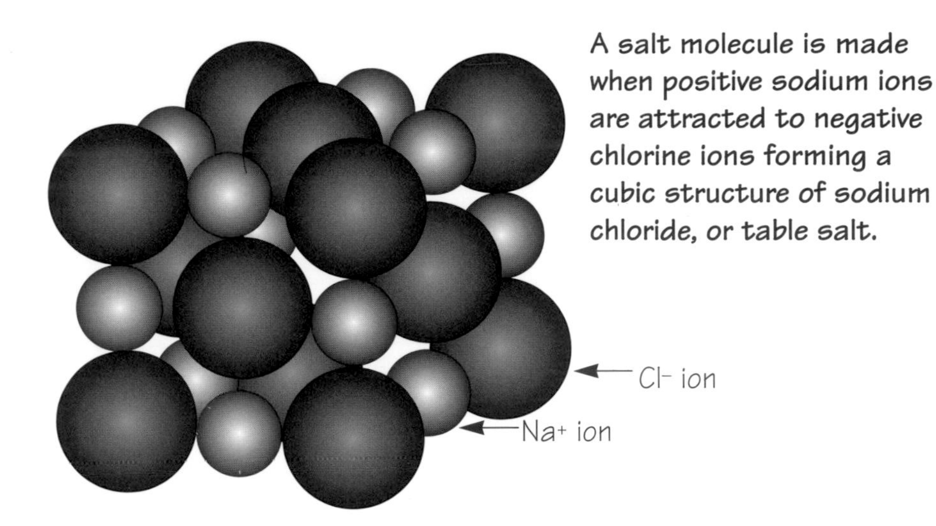

A salt molecule is made when positive sodium ions are attracted to negative chlorine ions forming a cubic structure of sodium chloride, or table salt.

other, salt crystals have a cubic structure as shown in the photograph on page 17. The biggest chunk of halite and the smallest grain of table salt both have the same cubic structure.

Chemical Salts

What most of us think of as "salt"—the white crystals we dispense from small shakers onto our food—is, in fact, only one kind of "salt." Chemically, a salt is the compound that results when a base, or alkaline substance, is neutralized by an acid. An acid is a compound that forms hydrogen ions (H+) when dissolved in water. A base is a compound that accepts such ions. When an acid and base react together completely, there are no ions left over. The acid has been neutralized by the base by forming a salt and water.

In nature, table salt is formed when sodium hydroxide—a base—neutralizes hydrochloric acid, yielding the salt and water.

$$NaOH + HCl \rightarrow NaCl + H_2O$$

When the water (H_2O) evaporates, the sodium chloride (NaCl) is left as a crystalline solid.

It seems as if all salts should be neither acidic nor basic, but, in fact, they may be either. If the acid being neutralized by a base is stronger than the base, the resulting salt is somewhat acidic. If the acid being neutralized is weaker, the salt is somewhat basic. However, NaCl is truly neutral. The negative chloride ions (Cl–) and the positive sodium ions (Na+) are equally strong.

All salts in solution or in molten form conduct electricity because they freely give up electrons. Electricity is the flow of electrons. Molten sodium chloride (it melts at 801°C; 1,474°F) is easily decomposed by electricity into positive sodium ions and negative chlorine ions. In electrolysis, the negative chloride ions are drawn to the anode, where they are oxidized (losing an electron) and form Cl_2, or chlorine gas. The gas bubbles up through the molten material. The cathode collects the sodium metal.

Sodium is still produced by electrolysis. Today, however, it is carried out in a special apparatus called a Downs Cell that captures the liquid metal sodium before it has a chance to react with water or air.

Grayish-white salt deposits in a Utah canyon formed as sediment that remained on the land when ancient seas evaporated, leaving behind salt-bearing sediment.

Mineral Salts

Much of the salt that eventually makes its way to our tables is found in large deposits all around the world. In some places these salt deposits are several thousand feet thick. It's unlikely that we'll ever run out of salt.

Sodium chloride that occurs deep underground is usually mined as if it were coal or a mineral ore. In the United States, about 85 percent of the table salt we use is mined from deep in the earth where deposits were formed millions of years ago. Some layers of salt are so far under the surface that they are inaccessible. However, where the rock over that salt is weak, the salt has sometimes risen upward to fill spaces in the rock. These salt-filled spaces, called salt domes, are often the most accessible sources of salt.

In other places around the world, water has seeped into a salt deposit and has not been able to evaporate, forming an underground salt, or brine, well. The thick, sludgy brine can be pumped up to the surface of the ground. If spread out in large, shallow pools, the water in the brine evaporates and salt crystals are left behind.

In a similar process, seawater is piped into large ponds where it is allowed to evaporate, leaving salt crystals. As the seawater gets thicker and thicker, different minerals crystallize at different times. Seawater can be "mined" for such minerals as salts of bromine (Br, element #35) and magnesium (Mg, #12), as well as sodium chloride.

Salt is also found in some of the huge, very flat areas called dry salt lakes, where bodies of water with a high salt content existed in the past. The vast flat regions of Bonneville Salt Flats in Utah are used for speed tests of jet-propelled cars. The area at Edwards Air Force Base in California where the space shuttles land is also a dry salt lake.

Ancient salt lakes evaporated to form flat terrain. The National Aeronautics and Space Administration (NASA) takes advantage of this flatness at Edwards Air Force Base by using it to land shuttles returning from space.

Why Salt Preserves Food

Food spoils, or rots, because various bacteria and molds act on it. These microorganisms (microscopic living creatures) are in the air around us at all times. However, most of them are unable to act on food if there's no moisture in it. For centuries, humans have used salt to preserve foods, especially meat and fish, for long periods of time. The salt actually "dries" the food by pulling water out of the food's cells. This drying-out process, which preserves the food for a certain period of time, happens through a process called osmosis.

Perhaps you have dropped Easter egg dye into a cup of hot water. You may have noticed that the dye spread out through the water without being stirred. The tendency of the molecules of one fluid to spread out evenly among the molecules of another fluid is called diffusion.

When a porous membrane (such as a cell wall) lies in

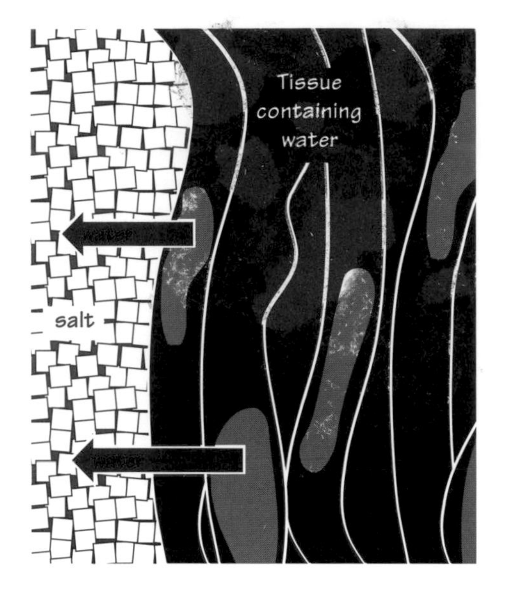

A coating of salt preserves meat and some other foods by drawing water out of the food tissues by the process of osmosis. The water moves out through cell membranes to try to balance the concentration of salt on each side of the membranes. With the water gone, bacteria cannot easily attack the food and make it spoil.

between two different fluids, the two fluids will try to mix evenly with each other. But cell walls are semipermeable, which means they will let some substances through and not others. In fact, a definition of osmosis is "diffusion through a semipermeable membrane."

Some foods can be preserved by coating them with salt, which draws the moisture out of the food. Usually salt-preserved food must have its water replaced by soaking the food before it can be cooked or eaten.

The substances within the cells of various organs of the body are kept balanced in much the same way. The fact that there are more molecules of a substance on one side of the cell membrane than on the other causes water to flow through the membrane. The water tries to bring equilibrium, or balance, to the solutions through osmosis.

Usually human beings will find water drinkable if it has a maximum sodium chloride content of 500 parts per million. An osmosis process is one of the ways that seawater can be made drinkable. Saltwater is pumped through a long tube made of semipermeable membranes. As it flows through the tubes, water molecules go through the small pores of the membrane and are collected on the outside of the tube, but the larger molecules of salt and other minerals remain behind in the tube. This method of obtaining drinkable water is used primarily in wealthy oil countries of the Middle East because it costs a great deal to run the pumps that make the process work.

Salt and Osmosis in Plants

Osmosis is responsible for water moving through the entire structure of a plant. Water in the soil enters the roots through cell walls because the substances on each side of the walls have different concentrations. Then the water moves from the first cells into the next ones, for the same reason. This continues until the water reaches the tubes that carry water upward into the main trunk or stem of the plant.

Above the ground, the water makes its way by the same process of osmosis into the leaves. From the leaves, the water is continually evaporating into the atmosphere. This movement of water through the cells by osmosis establishes an entire cycle of water through plants and their

Mangroves are among the few plants that can live in salty water. Their root systems often trap silt in the water, and the silt gradually adds to the land area of tropical islands.

environment. However, the presence of salt in water can prevent the needed flow of water through the plants. Very few plants can survive in a salty environment.

Salt in the Sea

About two-thirds of the solid material dissolved in the sea is sodium chloride. The actual amount of salt in seawater varies from place to place and from time to time. There is less salt in regions where icebergs float and melt than there is in parts of the same ocean where there are no icebergs. Icebergs contain frozen fresh water, which they release as they melt. There is also much less salt in seawater near the mouth of a river because the river's current carries fresh water into the ocean. However, deep water all over the world is equally salty.

The saltiest bodies of water on Earth are not its oceans. Some inland lakes have very high concentrations of salts. The salts just accumulate until, sometimes, few living things can survive in the water. The Great Salt Lake in Utah, for example, is about 12 percent sodium chloride and 3 percent other dissolved solids. The Dead Sea in the Middle East is even saltier. In fact, a layer of water about 110 meters (360 ft) deep in this body of water holds as much salt as it is possible for water to hold—it is saturated. Salt and some other minerals dissolved in the Dead Sea are collected by evaporating its water.

In the nineteenth century, some scientists thought they could figure out how old our planet was by measuring the amount of salt in ocean water at a given time and then measuring the salt content again at a later time. The difference would indicate how much sodium came into the ocean during the period between tests. The scientists thought that by measuring the amount of salt in the ocean at different times, they could figure out how old the ocean was. One figure derived that way was 90 million years—only several billion years off!

The problem is that sodium doesn't just keep accumulating. Instead, like most chemical elements in the Earth's crust, it is actually part of a continuous cycle of accumulation and dispersal. Some of the salt in the ocean settles to the bottom, becoming part of the ocean floor's sediment. The sediment gradually hardens and moves with the shifting of the huge plates making up the Earth's crust. These sediments become part of the continents, where they gradually reach the surface and become subject to weathering again. The salt molecules are washed away and enter the ocean once again.

Salt and Temperature Change

Lakes and rivers often freeze, but oceans almost never freeze, because sodium chloride freezes at a lower temperature (–1.9°C;

This semi-trailer is being loaded with salt to take to communities where the salt will be spread over icy roads.

28.6°F) than water freezes (0°C; 32°F). This same characteristic makes salt useful on icy roads and sidewalks. If you sprinkle an icy sidewalk with even a few crystals of sodium chloride (potassium and calcium chloride salts are also used), the ice melts around the spot where the crystal lands because the freezing temperature of the ice near the crystal has been lowered. The surface becomes pitted and rough, making the ice less slippery and safer to walk on.

On a larger scale, salt is often scattered on icy roads to melt ice or prevent water from freezing. However, much of that salt gets washed into the soil at the side of the road, polluting both the soil and the groundwater beneath it. Because of that pollution danger, many towns and cities have stopped salting their highways and are spreading sand instead. The sand provides something that a tire can grip, but it does not change the character of the ice.

On the other end of the temperature scale, salt also prevents water from boiling at 100°C (212°F). When salt is put in water being used for cooking—spaghetti, for example—the water boils at a higher temperature.

Salt in Soil

You may see a band of white crusty material at the seashore or beside a lake. This white crust is a mixture of a number of mineral salts that crystallized when the sun's heat evaporated the water in the lake.

Most water—even so-called pure rainwater—contains some salts. These salts can accumulate in soil, perhaps harming it. The salts may come from river water that has flooded over its banks or from the weathering of nearby rocks. But they also come when farmers irrigate their fields with salty water. The *occasional* use of water with a high content of sodium chloride and other salts doesn't hurt, because rain will wash the salts away.

But if farmers use such water too often, the salt builds up on the soil, gradually making it useless for growing crops.

Fresh water is found inside the Earth's crust in special kinds of rocky areas called aquifers. When we dig a well, we are digging down into an aquifer. Over the centuries, the water that people have removed by digging wells has been replaced by water seeping into the ground from rain.

In recent decades, however, more water has been removed from some aquifers than has been replaced. In some areas near the sea, an aquifer that no longer contains enough "pure" water will allow saltwater from the sea to seep in. People are now getting saltier water to drink than they have ever had before. If that water is used for irrigating crops, as it usually is, the soil could become too salty, too.

Any water flowing into Death Valley, California, quickly evaporates, leaving the salts the water carried behind as crystals that have piled up over many thousands of years.

BLOOD, SWEAT, AND TEARS

Active people need to replenish the supply of sodium in their bodies because sodium is lost through sweating.

About 0.15 percent of the human body by weight consists of sodium. The element in the body is mostly in the form of the salt sodium chloride, NaCl.

Sodium is a macromineral—one of the minerals needed by our bodies in major amounts. Other macrominerals are calcium potassium, magnesium, chlorine, phosphorus (P, element #15), and sulfur (S, #16). Sodium, calcium, potassium, and magnesium all function in the body as ions.

Electrolytes

Crystalline salt is formed by ionic bonds between sodium and chlorine. When salt is dissolved in water, those bonds break, releasing positive sodium ions and negative chloride ions into the water. The ions are now free to move and take on or release electrons. Such a solution will conduct electricity, which is the flow of electrons. Like the fluid used in electrolysis, fluids in our bodies that perform in this way, as well as the ions themselves, are called electrolytes.

About two-thirds of the fluids in the body are contained within cells, and about one-third is outside the cells. The fluid inside the cells gives the cells their shape and holds the various substances that make the cell function. Most fluid outside the cells fills the spaces between cells and between organs. This outside fluid is as important to the cells as the atmosphere is to us.

Two significant electrolytes in our bodies are sodium ions, Na+, and potassium ions, K+. Sodium ions are required by body cells to be in high concentration outside the cells and in low concentration within the cells. Potassium ions are the opposite—high concentration inside the cells and low outside. Some sodium ions also exist in muscle-cell fluids, but potassium ions are more important in muscles than sodium ions are.

Ions in the blood keep the body functioning properly. Electrolytes in the nervous system, for example, transmit minute electrical signals, causing nerve cells to signal muscle cells to move. Triggered by an electrical impulse, a nerve cell releases a chemical called acetylcholine. The chemical moves to a muscle fiber and prepares it to accept sodium and potassium ions. Those ions signal calcium ions to set up a complex chemical reaction that causes muscle fiber to contract.

Osmosis and Body Fluids

A great deal of water continually circulates through the body, especially in the liquid part of blood called plasma. The water does different jobs in different locations. Much of this flow happens because of the concentration of sodium chloride.

Blood itself is saline, meaning that there is salt in the plasma, the fluid part of the blood. As the blood moves through the

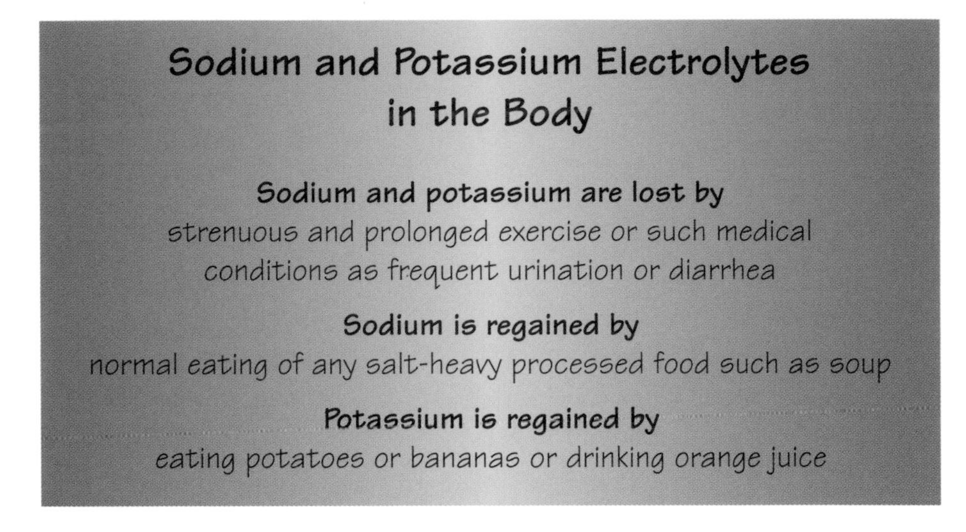

various organs of the body, various substances move into and out of the blood by osmosis.

For example, much more water passes through the kidneys than is released as urine. As the blood carries various waste substances into the kidneys, water flows through the various parts of the organ and back into the bloodstream by osmosis, leaving the waste substances to gather in one place. Along with some water, the waste is finally excreted as urine. Some salt is also excreted in the urine, which is one of the reasons why our diet must include salt to replenish that which is lost.

Not a Drop to Drink

There are many stories of people adrift on the sea in a lifeboat. They see water all around them, but if they drink it, they can die of thirst. If salt is needed to make our bodies work, why isn't it safe to drink saltwater?

Animals that live in the sea have special mechanisms for keeping the water balanced in their bodies instead of flowing out because of osmosis. Saltwater fish drink a lot of water to replace what they lose by osmosis. The salt content of the water they drink is passed out through their gills or in urine.

Humans, of course, don't have such mechanisms. We're not meant to drink saltwater. When saltwater hits the stomach, it has a different sodium content than the water in the cells of the stomach. Because of osmosis, water flows out of the cells instead of into them. The cells then actually start to dry out. So drinking saltwater makes us more thirsty instead of less thirsty. The salt content can also make us nauseated.

Sweat and Tears

Sweating, or perspiration, occurs when the body needs to get rid of heat. This doesn't happen by the sweat carrying actual heat out of the body. Instead, when the sweat evaporates from the skin, it cools the skin. Sweat is given off by sweat glands in the skin. The odor of sweat does not come from the sweat itself. Instead, it comes from the decomposition of the sweat by bacteria on the skin.

Most of the composition of sweat—about 98 percent—is water. The remainder consists of various chemicals, including quite a bit of sodium chloride. On a really hot day, or when a person has exercised a great deal, more salt can be lost through sweat than is healthy. People often take salt tablets when they're perspiring a lot in order to replenish the salt lost.

Our eyes are continually bathed in tears that protect the eyeballs from dust and other pollution and from drying out. Tears are primarily water that contains both sodium chloride and sodium bicarbonate, as well as some other substances.

Balancing Sodium

Animals may not understand the chemistry underlying the need of their bodies for sodium, but if they need salt, they go in search of it. Farmers and game wardens often put out blocks of salt, called salt licks, where cattle or deer will find them.

An average person who weighs about 68 kilograms (150 pounds) has approximately 100 grams (3.5 oz) of sodium in his or her body. As we've seen, some of that is continually given off in urine and sweat, and even in tears. This amount has to be replaced. A healthy person should get between 2 and 15 grams (.07 to .53 oz) of sodium chloride in the diet each day.

Because tears, sweat, and urine take sodium ions out of the body, we need to take in more sodium. However, if the body has more sodium than it can handle properly, fluid can build up in various tissues, making them swell. This is a condition called edema. But if a body has less sodium than it needs, the body's tissues can become dehydrated, or lose water.

Edema is the swelling of body tissues—especially the hands and feet—caused by fluid accumulation when there is too much sodium in the body.

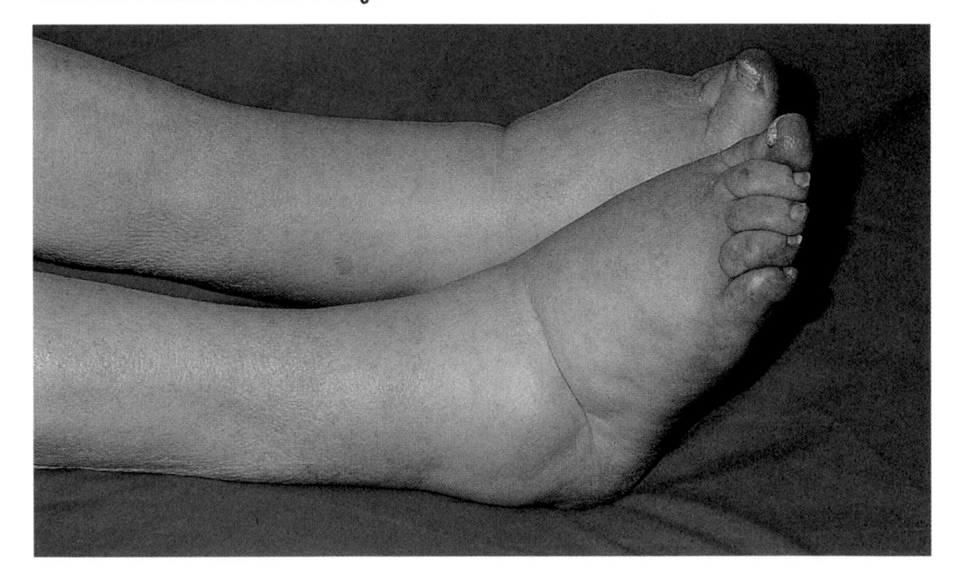

In recent years, a new way has appeared for people, especially athletes who sweat a great deal, to make sure they have enough electrolytes. The athletes consume so-called sport drinks. In 1965, the company that first put such drinks on the market tested them on the University of Florida football team. The team, called the Gators, happened to have a winning season that year, and the drink was named Gatorade® after them.

Under certain conditions, the body can't handle as much sodium as we normally get in our diet. This extra salt can cause high blood pressure and heart problems. Patients with those problems are often put on diets that restrict their salt intake to less than 1,000 milligrams of sodium a day. That may sound like a great deal of salt, but one teaspoon of table salt contains 2,440 milligrams of sodium. That's about the amount a healthy person needs to take in each day.

You may think you don't put much salt on your food, and maybe you don't. But take a look at the labels on some of the prepared foods in your pantry. A huge amount of salt is generally used in processing foods, primarily to enhance the flavor. A single serving of your favorite soup may contain almost half the salt you should have in a day.

Saline Solutions

Hospital patients are often given water containing salt directly into their veins. These patients have a needle inserted into a hand or arm, and the needle is connected to a tube running to a bag of fluid. The needle goes directly into the patient's vein and carries the fluid composed of water containing 0.9 percent sodium chloride—usually called a "normal saline" solution. This is the same percentage of sodium chloride found in normal body fluids. Called a saline drip, this fluid can have other substances added to it as needed—perhaps painkilling medicines or nutritious sugar if the patient can't eat.

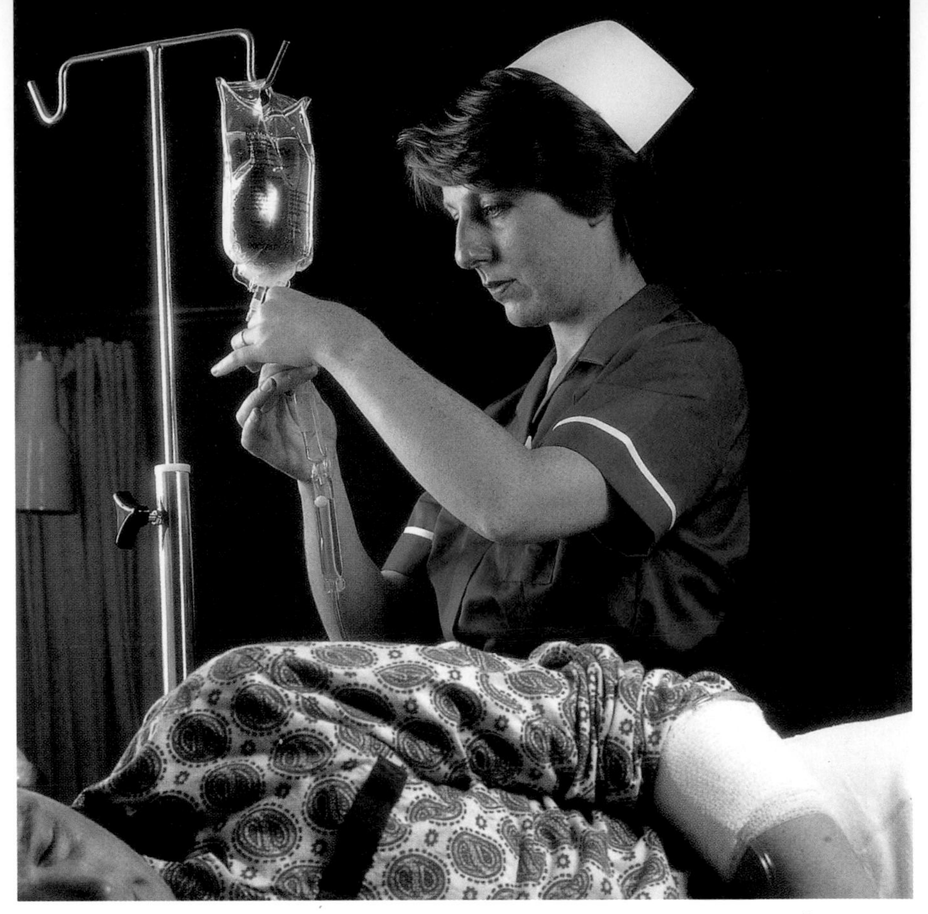

A nurse in a hospital checks the connections on a patient's saline drip after starting a new bag of saline solution.

Pentothal® (chemically called thiopental sodium), for example, is a drug that puts patient to sleep within seconds. If the supply going into the vein is turned off, the patient wakes up very quickly, although he or she generally sleeps for quite a while afterward. Pentothal is often delivered in a saline drip to a patient before an operation. The patient is already asleep before ether or another general anesthetic is used to keep the person asleep as long as the surgeon needs to work.

Radioactive sodium-24 can be injected into the blood circulation system through a saline drip. The movement of the isotope through the patient's circulatory system can be traced with a device that measures radioactivity. This technique is used especially for tracking the time it takes for blood to circulate throughout the body, indicating the health of the heart and circulatory system. This test also shows if there is a blockage in a vein or artery.

Sodium Bicarbonate in the Body

Sodium chloride is not the only sodium compound in our bodies. Sodium bicarbonate, $NaHCO_3$, for example, is present in blood plasma.

Cells burn sugars for energy. All burning (or oxidation) of organic compounds (which contain carbon—C, element #6) results in carbon dioxide (CO_2), which the body needs to remove.

Carbon dioxide leaves the cells and enters the bloodstream, where the CO_2 molecules link up with sodium and water, forming sodium bicarbonate. The sodium bicarbonate travels through the blood until it reaches the lungs. There, the carbon dioxide is released and exchanged for fresh oxygen. The sodium circulates back to the cells where it picks up more carbon dioxide.

Sodium bicarbonate molecules also play a role in the acid balance of the blood. A buffer is a chemical that prevents a solution from becoming either too acidic or too basic. Buffering by bicarbonate molecules keeps blood near its normal slightly basic level. Sodium bicarbonate can also play a role in the acid balance of the stomach. It serves as an antacid, a chemical that neutralizes excess acid. Hydrochloric acid (HCl) is released by glands in the stomach as part of digestion, but sometimes there is too much hydrochloric acid in the stomach. Sodium bicarbonate neutralizes hydrochloric acid.

When the sodium bicarbonate breaks up, it produces sodium ions. People who are on low-sodium diets to prevent edema should not take sodium-containing antacids often. They should use a calcium-containing antacid instead.

SOAP AND WATER

Water is very important to life. It is part of every living cell, and its movement through the body makes many organs work properly. Water is also an important part of daily living. It is used for recreation, drinking, and washing and in industry. But water doesn't always behave as we would like it to, especially when we want to clean with it. Also, most water has something in it other than plain H_2O molecules, and we sometimes have to try to clean up the water itself.

In colonial times, soap-making was often a project involving several neighbors. Each household contributed some of the ingredients.

When Water Doesn't Work

Most of the cleaning we do involves water. But water by itself is not very effective at getting rid of greasy dirt. Rushing water can push dirt off a surface, but most greasy dirt contains hydrocarbons, which are molecules made up of only hydrogen and carbon atoms. For example, if you got buttered popcorn on your shirt, the butter—which contains hydrocarbons—would soak into the fabric and resist water.

Water molecules, H_2O, have a tendency to cling together instead of penetrating dirt. If the water doesn't penetrate the greasy dirt, it's not going to wash the dirt away. It takes soap or detergent to help water do its job. Actually soap is a type of detergent, but today we use the word "detergent" to mean a soap that is produced synthetically.

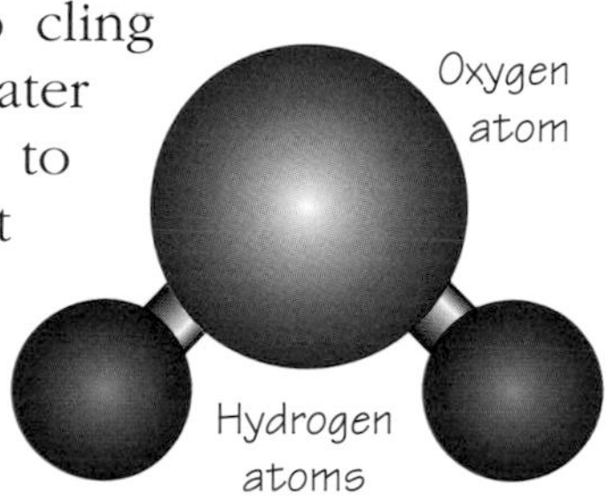

A water molecule consists of one oxygen atom and two hydrogen atoms.

Since ancient times, people have used soaps to help water work better at cleaning. The main ingredients in soaps are alkalis—usually sodium hydroxide (NaOH)—and animal fats. Stains from animal fats such as butter are very hard to remove from cloth without the use of soap. Strangely, animal fats are among the compounds used to fight dirt. What they do is make water "wetter."

Surface Tension

The tendency of water molecules to cling together is called surface tension. Perhaps you have experimented with seeing how much water you can put in a glass before the water's curved surface at the top of the glass breaks and the water spills. That's an experiment in surface tension.

The task of soaps and detergents is to reduce the surface tension of water—to make it "wetter"—so that the molecules will stop clinging together and will soak into fabric. A chemical that breaks up surface tension is called a surfactant (abbreviated from "surface-active agent"). Not all surfactants are detergents though. Some surfactants that also affect fluids are given as medicines to patients before surgery. Others are used to help break up oil on a beach after an oil spill.

The Two-Ended Molecule

Soap is an amazing chemical that was used to help clean things long before chemistry could explain how soap worked. Soap requires a combination of an acid substance and a base substance called an alkali. The acid content of soap is made up of organic (carbon-containing) molecules called fatty acids, which are types of carboxylic acids. The second ingredient of a soap is an alkaline material, which is usually sodium hydroxide (NaOH).

A typical soap is made by combining the sodium hydroxide with the fatty acid. Chemically, the reaction is written:

$$3NaOH + (C_{17}H_{35}CO_2)_3C_3H_5 \rightarrow$$
$$3C_{17}H_{35}CO_2Na + C_3H_5(OH)_3$$

As you can see from the chemical reaction, the sodium from the sodium hydroxide joins the fatty acids. The leftover by-product [$C_3H_5(OH)_3$] is called glycerol. When colonial women made soap, they also got a lotion (the glycerol) that they could rub into their skin.

The fatty acids and the alkali together form long molecules that have different characteristics at each end. Most of the molecule is a long chain of CH_2 groups. These long chains won't dissolve in water.

A typical soap molecule looks like this:

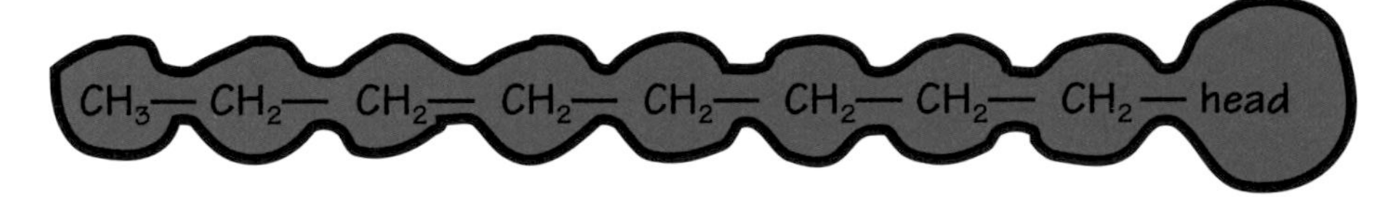

The head of a soap molecule is ionic. It is made up of a positive sodium ion attached to a negative carboxylic ion. This "head" is called hydrophilic, which means "water-loving." It refuses to mingle with hydrocarbon, or oily, substances but easily dissolves in water. The other end of the long chain is a tail of a CH_3 group. It is described as hydrophobic, or "water-hating." That molecule refuses to dissolve in water but is attracted to the greasy substances, usually hydrocarbons, which are the biggest problems in getting clothing clean.

The chemical structure of the head of a soap molecule

When the washing process starts, the hydrophilic head of the soap molecule decreases the surface tension of the water so that the water can be absorbed by the cloth. The hydrophobic tail attaches to the dirt molecules, pulling them out of the cloth.

Soap molecules then form a cluster with the hydrophilic heads on the outside. These spherical clusters are called micelles. Micelles are attracted to water molecules.

Each micelle encircles a tiny amount of dirt, holding it away from the clothing until the soap and the dirt are washed away by flowing water.

The more the dirt, water, and soap are moved around, the more fully micelles form all around the

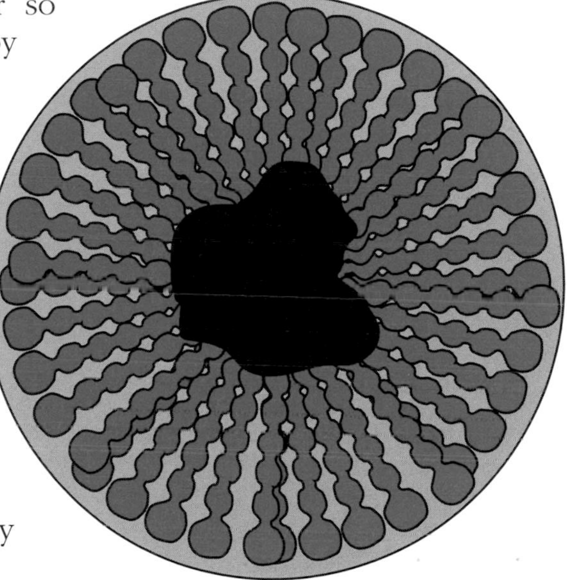

A greatly enlarged micelle, formed by soap molecules surrounding oily dirt

greasy dirt. That's why automatic washers have an agitator in them to move clothes around and why we get our hands cleaner if we rub them together

Making Soap

When people made their own soap, they saved the fat from animals they had butchered and the ashes from their wood fires. When they dissolved and boiled the ashes in water, they obtained potash. When the potash was cooked for a long time with the animal fat and then cooled—a process called saponification—the substance hardened into soap. The soap could be poured into a frame, cut into blocks, and shared with the neighbors. In some parts of the world, olive oil or coconut oil served the same purpose as animal fat.

In the late eighteenth century, a French chemist named Nicolas Leblanc developed a way to make sodium hydroxide and sodium carbonate out of common salt. The sodium hydroxide, also called caustic soda, could be used in making soap. For the first time, soap could be made and sold at prices common people could afford.

Detergents

It wasn't until the twentieth century that science moved beyond old-fashioned soaps and made the next step in cleaning. This was the invention of detergents using ingredients taken from petroleum. These ingredients have the ability to work in water that contains dissolved minerals, especially calcium and magnesium ions. Water containing these ions is described as "hard" water. When soaps are used in hard water, they do not form micelles and get washed away. Instead, the soap forms a nasty scum that doesn't rinse away easily. This scum forms because the negative ions in the soap bond with the positive mineral ions in the water, making scum instead of dissolving the dirt. A gray ring that forms around a bathtub is the scum from hard water.

Detergents were invented in the 1930s to overcome the problem of washing with hard water. Detergents work to wash clothes regardless of the hardness or temperature of the water. Detergents also contain ingredients that add a thin protective coating to washer parts to prevent them from rusting. Powdered detergents have an ingredient that keeps the detergent dry enough to pour properly.

Detergents aren't the answer to all cleaning problems, however. Chemists have not been able to make detergents that do a good job washing clothing but are gentle enough to use on the body. In addition, soap made for use on the body still does not work well in hard water.

To make water softer, some municipalities add a chemical called slaked lime, $Ca(OH)_2$, to the water. The slaked lime reacts with the calcium and magnesium in the water. A solid substance, or precipitate, forms and settles to the bottom of the water container, from which it can be removed. However, this process still leaves water that is harder than most people like. To solve the problem, many people have installed special devices called water softeners in their homes.

Softening the Water

The main job of a water-softening device is to replace the hard-water mineral ions with ions that don't prevent soap from working. This process is called an ion exchange. Water softeners substitute sodium ions (Na^+) for calcium and magnesium ions in hard water.

The main substance in an ion-exchange softener is a supply of granular or bead-like minerals called resins. In the early days of ion exchange, these resins were natural minerals called zeolites. Today, synthetic resins are used.

When the hard water goes through the resins, the resins remove the calcium and magnesium ions and exchange them for sodium ions. The resins are given new supplies of sodium ions by flushing brine, or saltwater, through them. That's why bags of salt must regularly be added to the water softener device.

A cutaway view of a home water softener

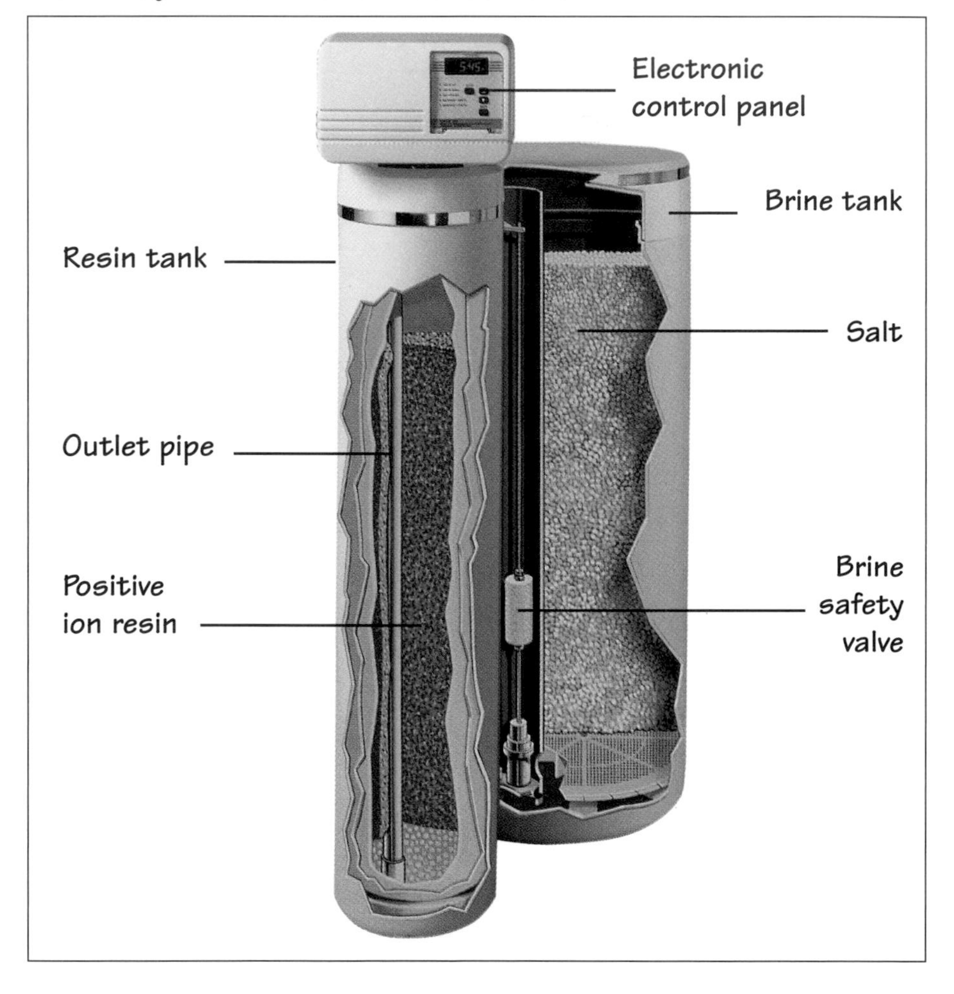

INDUSTRIAL USES OF SODIUM

Sodium hydroxide, or caustic soda, is probably the most important chemical base used by industry. In fact, it has been said that the chemical industry got its start when chemists learned to convert sodium chloride, NaCl, into sodium hydroxide, NaOH.

In the chemical industry, sodium hydroxide is still made from sodium chloride because of the great abundance of salt in the Earth. You have already seen the role that sodium hydroxide plays in soap, but it has many other uses as well.

Rayon threads being made from wood pulp that has been treated with sodium hydroxide

Caustic Soda

If sodium touches the skin, it reacts instantly with any moisture, changing into sodium hydroxide, NaOH, which burns the skin. In fact, this compound is so corrosive that it is the main ingredient used to clean drains. It has no other direct use at home, but it has many uses in industry.

In industrial quantities, sodium hydroxide is manufactured from sodium chloride by electrolysis. The chemical equation is written:

$$2NaCl + 2H_2O + \text{electricity} \rightarrow 2NaOH + Cl_2 + H_2$$

The process also produces diatomic (two-atom molecules) hydrogen gas and diatomic chlorine gas. Both gases are collected and used by various industries.

Sodium hydroxide is an important chemical in the paper industry. This is because of what it does to cellulose, the fibrous plant material that is used to make paper.

The cellulose in soft woods such as pine is the basic material used by the huge paper industry. To make use of cellulose, the material must first be dissolved and softened. Cellulose won't dissolve in many of the common solvents, such as water or alcohol, but it will dissolve in a solution of sodium hydroxide. The chemical also makes the fibers swell so that they are softer and easier to work with. It takes less energy to process paper pulp that has already been softened by NaOH.

Another product made from cellulose is cellophane, a plastic film used widely in packaging. Like paper, cellophane is produced from wood pulp. The pulp is treated first with sodium hydroxide, then with a series of other chemicals. The pulp turns into a thick fluid called viscose. As viscose is pressed out in thin layers, it is bathed in another sodium chemical—sodium sulfate, Na_2SO_4. These are just a few stages in the complex manufacture of this useful plastic. Viscose can also be filtered and spun into the useful threads called rayon.

About 1850, an Englishman named John Mercer invented a process for using sodium hydroxide to make cotton thread shrink and take on a glossy coating. Once the thread was "mercerized," it couldn't shrink again when the clothing made from it was washed. For many years the most common thread used for sewing was called mercerized cotton.

Olives are also treated with sodium hydroxide. Olives contain a chemical called a glucoside, which makes them taste naturally bitter. In preparing olives for bottling or canning, olives are treated several times by soaking them in a solution of sodium hydroxide. The sodium hydroxide reduces the bitter taste by neutralizing the glucoside. The olives are then washed and bottled in another sodium substance—salty water.

Trona and Glass

Not all salt deposits are sodium chloride. Seas with other concentrations of sodium chemicals have evaporated throughout the history of the earth, leaving other deposits behind. Probably the most important such deposit is trona. This soft mineral is a combination of sodium carbonate and sodium bicarbonate. It has the chemical formula $Na_2CO_3 \cdot NaHCO_3 \cdot 2H_2O$. Mined today primarily in Wyoming, it is the main source of the chemical sodium carbonate, Na_2CO_3, or soda ash, used in making glass.

Glass has been made for about 4,000 years. Since the beginning, its primary ingredient has been sand, which is silica, or silicon dioxide, SiO_2.

Molten soda-lime glass (at the left in the picture) fills molds to make bottles.

Most glass made today is used in bottles and windows. And most of that glass is of the variety called soda-lime glass. The most common formula for soda-lime glass is 72 percent silica, plus about 15 percent soda ash (sodium carbonate) and 9 percent lime (calcium oxide, CaO).

The beautiful and more expensive kind of soda glass is called crystal. It contains lead oxide instead of calcium oxide. (Lead is Pb, element #82.)

Amazing Sodium Bicarbonate

In the 1860s, Belgian chemist Ernest Solvay discovered a way to make sodium bicarbonate, $NaHCO_3$. He heated calcium carbonate, or limestone, to make it decompose into calcium oxide, or lime, and carbon dioxide, CO_2. The carbon dioxide was collected and bubbled into a solution of sodium chloride and ammonia, which is NH_3 in water. The resulting reaction produced sodium bicarbonate and ammonium chloride.

$$NaCl + CO_2 + H_2O + NH_3 \rightarrow NaHCO_3 + NH_4Cl$$

Sodium bicarbonate is a white powder that we often call baking soda. It has an amazing number of uses.

One of the most important uses of sodium bicarbonate is as leavening—an ingredient that makes a flour dough lighter. When mixed with flour and liquid, sodium bicarbonate changes to carbon dioxide, forming bubbles in the dough that make it lighter. A recipe containing baking soda must be cooked immediately, otherwise the carbon dioxide bubbles will disappear and so will the leavening effect.

Baking soda can be stirred into water and drunk to neutralize heartburn, which is caused by excess hydrochloric acid in the stomach. However, in neutralizing the acid, carbon dioxide is also made. This gas can exert pressure on the stomach, making the person feel very uncomfortable. In addition to carbon dioxide, the reaction also makes water and sodium chloride. The reaction is chemically written:

$$NaHCO_3 + HCl \rightarrow CO_2 + H_2O + NaCl$$

Baking soda is also useful for brushing teeth. It absorbs odors and is used as an abrasive for cleaning. People take advantage of the fact that baking soda absorbs odors by keeping some in the refrigerator, where it prevents one food from taking on the odor of another. Baking soda is also useful in a closet, placed near bad-smelling shoes.

Cooks often add a pinch of baking soda to pots of vegetables. The compound helps the vegetables retain their vivid green color, making them look more appetizing. However, baking soda reduces the vitamin C content of the food.

Marble monuments such as this one at Valley Forge National Historical Park are being cleaned with an environmentally safe process called soda-blasting. Crystals containing sodium bicarbonate are blasted at the surfaces to be cleaned.

An example of sodium bicarbonate's versatility is described in the *Woman's Day Encyclopedia of Cookery:* "Take baking soda on a fishing trip? Certainly! It will brighten up those dull or rusted lures and hooks, rid your hands of fishy odors, take the soreness out of an accidental sunburn, and soothe the itch or sting of bug bites."

Baking soda is also an ingredient in another leavening agent—baking powder. The other two ingredients in baking powder are an acid salt, with which the sodium bicarbonate reacts to produce carbon dioxide, and flour or cornstarch, which is added to prevent the other ingredients from absorbing moisture and forming clumps. A certain type of baking powder called "double-acting" baking powder contains sodium aluminum sulfate (referred to as "SAS") and calcium acid phosphate. These two ingredients work at different times in the cooking process so that carbon dioxide is given off immediately, and then again later while the recipe is baking.

More Sodium in Foods

People who have been told by their physician to cut down on their use of sodium may think they can achieve their goal by not adding salt to the foods they eat. But to truly cut down on sodium intake, people should be aware of the many sodium compounds used in food processing.

Sodium phosphate (Na_3PO_4) is added to packaged foods to keep the acid-base balance correct. Monosodium glutamate, called MSG, is used to strengthen the natural flavors in food. Sodium benzoate is added to help preserve acidic foods, such as grapefruit juice, when they are packaged in cans or bottles that don't have to be refrigerated. The chemical prevents the growth of microorganisms in an acidic environment. Sodium nitrite ($NaNO_2$) does the same thing in meat products, and sodium propionate helps preserve baked goods.

Sodium ascorbate is often added to foods. It is the sodium salt of vitamin C, or ascorbic acid. Vitamin C is an important compound for maintaining and strengthening body tissues. Some people swear that large doses of vitamin C can prevent serious colds.

Several other sodium compounds are used in dried products that must be cooked in water, such as macaroni for macaroni and cheese. Some sodium compounds give products a meaty flavor, even though there may be little or no meat in them. Disodium guanylate and disodium inosinate, for example, are laboratory chemicals that are also found naturally in meats. Adding these chemicals to some products gives them a meatlike flavor.

Other Uses of Sodium Compounds

Sodium compounds are used as bleaches, chemicals that remove color from other substances. Such bleaches are called oxidizers because the oxygen in the compounds reacts with the color-making chemicals and bleaches the colors out. Sodium chlorate ($NaClO_3$), for example, is used to bleach wood pulp.

This vat of wood pulp to be used in making paper has been bleached white with a sodium chlorate bleach.

Common laundry bleaches used at home are often made of sodium perborate, which works best in hot water.

Until recently, metallic sodium was used in manufacturing tetraethyl lead, $(C_2H_5)_4Pb$. This compound was added to gasoline to keep engines from "knocking" (making a noise when the fuel ignites in a cylinder). However, the lead that was given off into the air from internal combustion engines was a major source of air pollution and a health hazard. A law was passed requiring the use of only unleaded gasoline. Car manufacturers had to find other ways to prevent engine knock.

Sodium nitrate, $NaNO_3$, is one of the ingredients in dynamite. Called Chile saltpeter because the world's largest supply of the natural mineral is in Chile, this compound is also a source of many nitrogen chemicals.

A type of fire extinguisher used in fighting electrical fires has sodium bicarbonate as its primary ingredient. The heat of the fire decomposes the powdered chemical, making carbon dioxide, water, and a salt form:

$$2NaHCO_3 + heat \rightarrow Na_2CO_3 + H_2O + CO_2$$

The carbon dioxide blankets the fire, preventing oxygen from reaching the flames. This device is sometimes called a dry fire extinguisher because there is no liquid involved.

A special kind of streetlight called a sodium-vapor lamp has been in use since the 1930s. It gives off light because as an electric charge jumps the gap (makes an arc) between electrodes it vaporizes a tiny amount of sodium in between, giving off a yellowish light. This type of light looks nice outdoors, especially in narrow, old streets, but it makes people's skin look greenish. Sodium-vapor light also makes it difficult to distinguish the colors of things such as cars.

The mineral borax contains the element boron (B, element #5). When borax is added to glass, it turns into a tough, heat-

Deposits of borax found near the surface of Earth's crust can be removed from open mines by heavy machinery. The white rock of the deposit (shown in the background) is crushed to a fine powder, which dissolves in water. When the water is evaporated, crystals of borax form.

resistant glass called borosilicate glass. This kind of glass is sometimes used to make heat-resistant household bowls and cookware. Borax is also added to plant fertilizers because of some plants' need for boron.

Borax usually is found in or near salt lakes. It also occurs within deposits of other minerals, especially ulexite and kernite. Because ulexite contains borax, it was one of the minerals prospectors were seeking when they discovered Death Valley in California. Kernite, found in California's Mojave Desert, is one of the prime sources of boron.

A SODIUM CATALOG

A laser beam aimed at the atmosphere's sodium layer is visible in the night sky.

Twinkling Stars

Turbulence, or motion, in the atmosphere—even when we can't see it—limits the use of telescopes on Earth's surface. Light from the stars is distorted by turbulence in the atmosphere. The turbulence changes what the astronomical instruments are apparently seeing. Even without instruments, a person looking at the night sky sees stars "twinkle" because of atmospheric turbulence. That's why photos taken using the Hubble Space Telescope, which is on a satellite above the atmosphere, are so much better than those taken from Earth.

Astronomers have long used specific stars, called guide stars, to help them establish basic information about the universe. But such stars are not always available in the section of the sky where the astronomers might want to work. In recent years, astronomers have created artificial guide stars by reflecting laser light off a layer of sodium atoms in the atmosphere. This layer of uncharged sodium atoms is approximately between 75 and 90 kilometers (45 and 55 mi) in the upper atmosphere.

When astronomers direct their laser at that sodium layer, the light is reflected back. The scientists measure how the laser light has been changed by turbulence in the atmosphere. The astronomers can then apply those changes to the light from the star they are studying and figure out what the actual star is like.

Hale-Bopp's Sodium Tail

In 1997, the comet called Hale-Bopp was visible in the night sky for several months. It was already known that comets have long dust tails that form when ice crystals from these visitors to the solar system thaw as they near the sun. The tails are visible because the solid particles in them reflect sunlight. Comets are also known to have a usually invisible tail of ions that form from the molecules of the icy nucleus. They also have sodium atoms near the nucleus—the solid icy head of the comet.

But Hale-Bopp gave astronomers a surprise when their instruments revealed in April 1997 that the comet had an additional tail of neutral, or non-ionized, sodium atoms. This unexpected tail, which was much thicker than comet tails usually are, glowed with the same yellowish tint that sodium shows in a laboratory test.

The tail grew to be about 600,000 kilometers (375,000 mi) wide and 50 million kilometers (31 million mi) long. But it did not follow the same path as the regular ion tail of the comet, which is affected by the solar wind. Consisting of fast-moving atomic particles emitted by the sun, the solar wind did not appear to affect the sodium tail.

People on Earth could see the Hale-Bopp comet's normal dust trail as well as its bluish trail of ionized molecules. Special instruments were needed to discover the third tail of uncharged sodium ions.

It will take a long time—and perhaps the appearance of a similar comet—before astronomers know why Hale-Bopp has this peculiar tail. Initial speculation suggested that perhaps the icy core itself consisted of a sodium-bearing material that broke up as the comet moved through the heavens.

Therapeutic Salt

In the past, many people suffered from a disfiguring condition called goiter. In this condition, the thyroid gland in the neck becomes enlarged due to lack of the element iodine (I, element #53) in the diet. Goiter generally occurred in regions far from the sea where people did not get much of this element, which occurs naturally in foods taken from the sea.

This sometimes fatal condition can be prevented in a simple way. A tiny amount (only 10 parts per million) of sodium iodide (NaI) is added to common table salt. The NaI does not change the flavor of salt, but regular use of iodized salt has turned goiter into a thing of the past.

Absorbing Sodium Nitrites

The name "nitrite" often refers to a chemical used in preserving pork and some other meats. Nitrite is actually sodium nitrite ($NaNO_2$). In addition to preserving meat, this chemical gives meats an acceptable "meat-like" color.

As people have become more aware of the chemicals they eat, some have become concerned about the use of sodium nitrites and sodium nitrates ($NaNO_3$) in food. You may already know when you are reacting to sodium nitrates, because some people—especially those subject to the severe, long-lasting headaches called migraines—can get "hot-dog headaches" from the nitrates used in preserving cold cuts and hot dogs.

Nitrates turn into nitrites. This fact may create problems even more severe than headaches. It is possible for sodium nitrite to react with some amino acids (molecules making up proteins) in the body to form nitrosamines. These are among the most powerful carcinogens (cancer-causing materials) known.

The use of sodium nitrites has not been banned—as carcinogens often are—for a very good reason. The use of $NaNO_2$ prevents the growth of the bacterium *Clostridium botulinum*, which causes botulism, the most deadly food poison.

There is the additional problem that not all sodium nitrites in our bodies come from compounds that are added to food. Indeed, two-thirds of the nitrites in our bodies occur when bacteria in our digestive systems convert naturally occurring sodium nitrate into nitrites. Sodium nitrate also occurs in many vegetables and other foods.

What can we do about the nitrosamines? Not much yet, but the addition of vitamin C to the diet is known to reduce the production of nitrosamines by the body. Also, nitrosamines are associated with cigarettes, beer, cheese, smoked fish, and various "instant" products.

Why 32 Degrees?

Gabriel Daniel Fahrenheit, a German physicist, developed a reliable thermometer in the early 1700s. He decided it would be convenient if the most used part of the measurement scale were divided into 100 units with the temperature of the human body being 100 degrees at the top and the temperature of freezing water being 0 degrees at the bottom.

Fahrenheit did not want people using his thermometer to have to deal with negative numbers. He lowered the freezing point of water by adding salt (NaCl) to the water before marking that point as O°. Later, the freezing point of pure water was set at 32° Fahrenheit. The average temperature of the human body turned out to be not 100°F but 98.6°F.

Twenty years later, Swedish scientist Anders Celsius began using a temperature scale that had pure water freezing at 0°. Most of the world now uses the Celsius scale.

Thirsty Salt

Salt is hygroscopic—it pulls water out of the atmosphere. During a hot, humid summer you may have noticed that the saltshaker on your table will not work properly. The salt has probably absorbed so much moisture that it forms clumps that won't pass through the shaker's holes. If you put a few rice grains in your saltshaker, the salt will soon start pouring again because rice draws moisture out of the salt.

Keeping Products Dry

Perhaps you have bought a new VCR or other piece of electronic equipment lately. In the box, you probably found a small packet that said "silica gel" on it. This packet contains a chemical that keeps moisture from harming the contents of the package during storage. Silica gel has the simple formula, SiO_2. But it's made by using a sodium compound called sodium silicate, Na_4SiO_4. When the silicate is added to water and an acid is added to the mix, the result is SiO_2, which can be dried into a powder.

The powder in the package does not absorb water, as you might suppose. Instead, it *ad*sorbs, meaning that the powder granules attract water molecules and hold them to their surface. And because it is a powder, there is a great deal of surface area in a small measure of silica gel to adsorb water vapor.

Preventing Cavities

The enamel of our teeth is made of a complex chemical called hydroxyapatite, which is a combination of calcium phosphate and other compounds. In the mid-twentieth century, it was discovered that hydroxyapatite could be made stronger by replacing part of the formula (the hydroxide ions, OH^-) with fluoride ions (F^-). (Fluorine is F, element #9.) This was done by adding sodium fluoride (NaF) to water and toothpaste.

Oddly, highly concentrated sodium fluoride is extremely poisonous, but an amount not exceeding 1 part per million is added to the water in many communities. This tiny amount has given most people stronger teeth, making frequent cavities almost a thing of the past.

In a Nuclear Reactor

Pure sodium is an excellent heat conductor. If kept in the closed pipes of a nuclear reactor where air can't reach it, pure sodium can be used to conduct heat away from the reactor core. However, some radioactive sodium isotopes are produced from stable sodium-23 when it is used this way. Radioactive sodium-22 or sodium-24 may get into the water that is used to help cool the reactor. And polluted water can harm living things.

Sodium in Brief

Name: sodium, from the word soda
Symbol: Na, from *natrium*, meaning "metal of soda"
Discoverer: Known since ancient times but isolated by English chemist Humphry Davy in 1807
Atomic number: 11
Atomic weight: 22.99
Electrons in the shells: 2, 8, 1
Group: 1, the alkalis; other elements in Group 1 with 1 electron in the outer shell include lithium, potassium, rubidium, cesium, and francium
Usual characteristics: soft metal that floats on water and reacts readily with other elements
Density (mass per unit volume): 0.96 gram per cubic centimeter
Melting point (freezing point): 97.83°C (208°F)
Boiling point (liquefication point): 882.9°C (1,621°F)
Abundance:
  Universe: 7th most common, but sodium and several other elements together make up only 1.1%
  Earth: sodium is one of several elements that together make up only 1.4%
  Earth's crust: 6th most abundant (2.6%). Sodium oxide (Na_2O) is the sixth most common compound in Earth's crust
  Human body: 0.1% in blood and other fluids
Stable isotopes: only one, sodium-23
Radioactive isotopes: Na-19 through Na-22 and Na-24 through Na-31

Glossary

acid: definitions vary, but basically an acid is a corrosive substance that gives up a positive hydrogen ion, H^+, equal to a proton when dissolved in water; indicates less than 7 on the pH scale because of its large number of hydrogen ions

alkali: a substance, such as an hydroxide or carbonate of an alkali metal, that when dissolved in water causes an increase in the hydroxide ion (OH^-) concentration, thus forming a basic solution.

anion: an ion with a negative charge

atom: the smallest amount of an element that exhibits the properties of the element, consisting of protons, electrons, and (usually) neutrons

base: a substance that accepts a hydrogen ion, H^+, when dissolved in water; indicates higher than 7 on the pH scale because of its small number of hydrogen ions

boiling point: the temperature at which a liquid at normal pressure evaporates into a gas, or a solid changes directly (sublimes) into a gas; also, the temperature at which a gas condenses into a liquid or solid

bond: the attractive force linking atoms together in a molecule or crystal

cation: an ion with a positive charge

chemical reaction: a transformation or change in a substance involving the electrons of the chemical elements making up the substance

combustion: burning, or rapid combination of a substance with oxygen, usually producing heat and light

compound: a substance formed by two or more chemical elements bound together by chemical means

covalent bond: a link between two atoms made by the atoms sharing electrons

crystal: a solid substance in which the atoms are arranged in three-dimensional patterns that create smooth outer surfaces, or faces

decompose: to break down a substance into its components

density: the amount of material in a given volume, or space; mass per unit volume; often stated as grams per cubic centimeter (g/cm^3)

diatomic: made up of two atoms

dissolve: to spread evenly throughout the volume of another substance

electrode: a device such as a metal plate that conducts electrons into or out of a solution or battery

electrolysis: the decomposition of a substance by electricity

electrolyte: a substance that when dissolved in water or when liquified conducts electricity

element: a substance that cannot be split chemically into simpler substances that maintain the same characteristics. Each of the 103 naturally occurring chemical elements is made up of atoms of the same kind.

enzyme: one of the many complex proteins that act as biological catalysts in the body

evaporate: to change from a liquid to a gas

gas: a state of matter in which the atoms or molecules move freely, matching the shape and volume of the container holding it

group: a vertical column of elements in the Periodic Table, with each element having similar physical and chemical characteristics; also called chemical family

half-life: the period of time required for half of a radioactive element to decay

hormone: any of various secretions of the endocrine glands that control different functions of the body, especially at the cellular level

hydrocarbon: a compound made of only carbon and hydrogen

inorganic: not containing carbon

ion: an atom or molecule that has acquired an electric charge by gaining or losing one or more electrons

ionic bond: a link between two atoms made by one atom taking one or more electrons from the other, giving the two atoms opposite electrical charges, which hold them together

isotope: an atom with a different number of neutrons in its nucleus than other atoms of the same element

mass number: the total of protons and neutrons in the nucleus of an atom

melting point: the temperature at which a solid becomes a liquid, or a liquid changes to a solid

metal: a chemical element that conducts electricity, usually shines, or reflects light, is dense, and can be shaped. About three-quarters of the naturally occurring elements are metals.

metalloid: a chemical element that has some characteristics of a metal and some of a nonmetal; includes some elements in Groups 13 through 17 in the Periodic Table

molecule: the smallest amount of a substance that has the characteristics of the substance and consists of two or more atoms

neutral: 1) having neither acidic or basic properties; 2) having no electrical charge

neutron: a subatomic particle within the nucleus of all atoms except hydrogen; has no electric charge

nonmetal: a chemical element that does not conduct electricity, is not dense, and is too brittle to be worked. Nonmetals easily form ions, and they include some elements in Groups 14 through 17 and all of Group 18 in the Periodic Table.

nucleus: 1) the central part of an atom, which has a positive electrical charge from its one or more protons; the nuclei of all atoms except hydrogen also include electrically neutral neutrons; 2) the central portion of most living cells that controls the activities of the cells and contains the genetic material

organic: containing carbon

oxidation: the loss of electrons during a chemical reaction; need not necessarily involve the element oxygen

pH: a measure of the acidity of a substance, on a scale of 0 to 14, with 7 being neutral. pH stands for "potential of hydrogen."

photosynthesis: in green plants, the process by which carbon dioxide and water, in the presence of light, are turned into sugars

plastic: any material that can be shaped, especially synthetic substances produced from petroleum

pressure: the force exerted by an object divided by the area over which the force is exerted. The air at sea level exerts a pressure of 14.7 pounds per square inch (1013 millibars), also called atmospheric pressure.

protein: a complex biological chemical made by the linking of many amino acids

proton: a subatomic particle within the nucleus of all atoms; has a positive electric charge

radical: an atom or molecule that contains an unpaired electron

radioactive: of an atom, spontaneously emitting high-energy particles

reduction: the gain of electrons, which occurs in conjunction with oxidation

respiration: the process of taking in oxygen and giving off carbon dioxide

salt: any compound that, with water, results from the neutralization of an acid by a base. In common usage, sodium chloride (table salt).

shell: a region surrounding the nucleus of an atom in which one or more electrons can occur. The inner shell can hold a maximum of two electrons; others may hold eight or more. If an atom's outer, or valence, shell does not hold its maximum number of electrons, the atom is subject to chemical reactions.

solid: a state of matter in which the shape of the collection of atoms or molecules does not depend on the container

solution: a mixture in which one substance is evenly distributed throughout another

sublime: to change directly from a solid to a gas without becoming a liquid first

synthetic: created in a laboratory instead of occurring naturally

triple bond: the sharing of three pairs of electrons between two atoms in a molecule

ultraviolet: electromagnetic radiation that has a wavelength shorter than visible light

valence electron: an electron located in the outer shell of an atom, available to participate in chemical reactions

For Further Information

BOOKS

Atkins, P. W. *The Periodic Kingdom: A Journey into the Land of the Chemical Elements.* NY: Basic Books, 1995

Heiserman, David L. *Exploring Chemical Elements and Their Compounds.* Blue Ridge Summit, PA: Tab Books, 1992

Hoffman, Roald, and Vivian Torrence. *Chemistry Imagined: Reflections on Science.* Washington, DC: Smithsonian Institution Press, 1993

Newton, David E. *Chemical Elements.* Venture Books. Danbury, CT: Franklin Watts, 1994

Yount, Lisa. *Antoine Lavoisier: Founder of Modern Chemistry.* "Great Minds of Science" series. Springfield, NJ: Enslow Publishers, 1997

CD-ROM

Discover the Elements: The Interactive Periodic Table of the Chemical Elements. Paradigm Interactive, Greensboro, NC, 1995

INTERNET SITES

Note that useful sites on the Internet can change and even disappear. If the following site addresses do not work, use a search engine that you find useful, such as:
Yahoo:

http://www.yahoo.com

or AltaVista:

http://altavista.digital.com

A very thorough listing of the major characteristics, uses, and compounds of all the chemical elements can be found at a site called WebElements:

http://www.shef.ac.uk/~chem/web-elements/

A Canadian site on the Nature of the Environment includes a large section on the elements in the various Earth systems:

http://www.cent.org/geo12/geo12/htm

Colored photos of various molecules, cells, and biological systems can be viewed at:

http://www.clarityconnect.com/webpages/-cramer/PictureIt/welcome.htm

Many subjects are covered on WWW Virtual Library. It also includes a useful collection of links to other sites:

http://www.earthsystems.org/Environment/shtml

INDEX